INDUSTRIAL CHEMISTRY OF COMMON MATERIALS

A Handbook of Brief Introduction of How Chemical Materials are Made

Kingsley Augustine

Table of Content

Table of Content 2

Development of the Chemical Industry................................4

Important Raw Materials..5

Divisions of the Chemical Industry 8

Plastics 11

Metallurgy...14

Pharmaceuticals..15

Glass...16

Ceramics...19

Paints 20

Cement..21

Soaps and Detergents...23

Food and Drinks..24

Economics of Industrial Processes....................................25

Importance of Industrial Chemistry...................................27

Cleaning up Wastes 30

Development of the Chemical Industry

The development of the chemical industry had important effects on chemistry. Soon after the Middle Ages, factories supplying soap, alum, acids and other basic chemicals were set up. As society became complex, the need to develop and produce important chemicals grew. Chemists set to work to fulfill these needs. Wars also served as a force in accelerating these developments.

In 1749, the Lead Chamber process for the commercial manufacture of tetraoxosulphate(VI) acid was developed. This initiated a whole new range of other chemical industries. Later, in 1791, the Leblanc process for the production of trioxocarbonate(IV) was invented. Extensive industrial chemical research, however, was not in full swing until 1856 when William Perkin discovered the first aniline dye. Then, research by teams of scientists became the norm in chemical industry. The Haber process for the manufacture of ammonia was developed under such conditions. This led to an increase in the demand for trained chemists.

Important Raw Materials

In order find out more about the various chemical industries, it is a good idea to visit such industries sited in your community. You will be able to see the raw materials used and how they are converted into the various products that are in great demand by society. The important raw materials are

- Coal
- natural gas,
- petroleum,
- limestone,
- salt or sodium chloride,
- metallic ores,
- sulphur
- and air.

All except air are mined from the earth's crust. Salt is obtained mainly from sea water. Many other raw materials are derived from these main raw materials. Chemicals derived from coal are by-products of the process by which bituminous coal is converted into coke. The gas and vapors are led to special equipment to yield fuel gas, ammonia and coal tar. Coal tar contains a number of chemicals such as benzene, toluene, phenol, naphthalene and cresols. Limestone rocks, caves and cliffs exist in many parts of the world. From limestone we get lime or calcium oxide which is used for making cement and concrete. Salt or sodium chloride occurs as rock salt in underground deposits and in sea water. Electrolysis of sodium chloride produces chlorine. The other main product is sodium hydroxide which is used in the manufacture of soap and textile, and in petroleum refining. The chlorine produced can be made to react with hydrocarbons obtained from fossil fuels. These chlorohydrocarbons are versatile raw materials. They are used in the making of synthetic rubber, paint removers, refrigerants and several types of Insecticides, and in dry cleaning. Sulphur occurs naturally in some places. It is also obtained as a by-product when sulphide ores of copper or zinc are treated. Most of the sulphur is converted into tetraoxosulphate(VI) acid. This acid is manufactured in much greater quantities than any other chemical. It is used in many chemical processes and in the manufacture of fertilizers, paints, dyes, explosives, synthetic fibers and accumulators.

Important metals such as iron, aluminum and copper are extracted from their respective mineral ores. This has given rise to the metal extraction industries in countries where the ores are available. The metals are used mainly in the construction of buildings and bridges, and in making machines, vehicles, ships and aircraft. Air is the chief source of oxygen and nitrogen. Nitrogen is used in the manufacture of ammonia by the Haber process. Ammonia is used mainly in the production of fertilizers and trioxonitrate(V) acid.

Divisions of the Chemical Industry

The chemical industry is founded on a wide variety of raw materials. Amongst the most important are coal, molasses, salt, mineral ores, water, air and fats and oils of animal and plant origin. Petroleum is a fairly recent addition to the list. Since the chemical industry produces a variety of products, it is useful to classify the industry on the basis of its products. We have chemical industries which produce

- basic chemicals such as acids, alkalis, salts and organic chemicals;
- chemical products that are used in the manufacture of artificial fibers and plastics;
- chemical products used as starting materials in other industries that manufacture substances such as paints, fertilizers and explosives;
- and chemical products such as cosmetics, drugs and soap for our personal use.

The chemical industry is now defined as one that uses chemistry to make chemicals from other chemical substances.

Heavy Chemicals

The most useful basic chemical that the chemical industry makes is tetraoxosulphate(VI) acid. It is used in many chemical processes - from making fertilizers to cleansing (pickling) steel. Sodium hydroxide for making soap and other things is also produced in large quantities.

Ammonia is another chemical that is manufactured in large quantities. It is used to make fertilizers as well as explosives. These and other chemicals are called heavy chemicals because they are made in very large quantities.

Fine chemicals
Drugs and chemicals produced in relatively small amounts are referred to as fine chemicals. These are also produced to a very high degree of purity. The fertilizer industry is an important one. Fertilizers are needed to increase crop yield to feed the world's rapidly increasing population. Three important elements have to be provided in suitable form in fertilizers. They are nitrogen, phosphorus and potassium (commonly referred to as N.P.K.). Nitrogen is supplied mainly in the form of ammonia which in turn is obtained from fossil fuels. The ammonium compounds used are ammonium trioxonitrate(V) and ammonium tetraoxophosphate(V). Sodium trioxonitrate(V) is also used as a source of nitrogen. Sometimes, ammonium tetraoxosulphate(VI), recovered as a by-product from the conversion of coal into coke, is also used. Phosphorus is obtained in nature as rock phosphate. It has to be treated with tetraoxosulphate(VI) acid to get it into a form that can be used in fertilizers. The phosphorus can be converted into superphosphates, triple superphosphates and tetraoxophosphate(V) acid. This acid can then be converted into ammonium tetraoxophosphate(V). Potassium, in the form of its several salts, is mined from

underground deposits. Potassium compounds, nitrogen compounds and phosphorus compounds are mixed together in most of the fertilizers used nowadays. The proportions in which they are mixed are usually given as the N.P.K. values. In many cases, small quantities of trace elements (such as zinc, boron, copper and molybdenum) are added, depending on their need in certain regions.

Plastics

The modern plastics industry developed to a great extent out of a search for a substitute for rubber during World War II. In the nineteenth and early twentieth century, the four main sources of raw materials for the plastics and synthetic organic chemical industries were coal, limestone, cellulose and molasses.

Coal yielded:
- benzene for the synthesis of poly(phenylethene), commonly known as polystyrene;
- phenol, methanal and urea for the synthesis of a range of thermosetting plastics.

Limestone, when heated with coal produced calcium carbide, which yielded ethyne under suitable treatment. The ethyne is used in the manufacture of poly(chloroethene), commonly known as polyvinyl chloride.

Molasses yielded ethanol from which ethene is produced. The ethene is used to produce a range of poly(ethene) plastics.

Nowadays, starting raw materials are obtained by the fractional distillation of crude oil or petroleum. The plastics are manufactured in their raw form as granules or powder. These are then exported to fabrication plants to be manufactured into various products. Most plastic processing methods involve heating the raw material (in the form of granules or powder) until it softens, shaping the softened material and then setting it. Two important processes used are injection molding and extrusion. In injection molding, the plastic granules are heated until they are soft enough to flow. The softened plastic is then forced under pressure into a closed mould cavity where it can cool to form the shape of the mould. Since the cavity is filled under pressure, every detail of the mould would be faithfully reproduced in the finished part. In the extrusion process, a continuous flow of molten plastic is forced through a shaped opening called a die, and the plastic is then cooled. Depending on the shape of the opening, many different parts can be made. If the opening is like a ring, a tube or pipe can be made. If the opening is a long thin slit, a sheet may be formed. One application of extrusion is in the plastic covering of wires and cables for insulation purposes. Another adaptation is in extrusion coating. Paper, fiberboard and aluminum foil are coated with polymers by the direct extrusion process. The plastic coating enhances the packaging properties. For example, plastic coated paper is impermeable to water and oil and is hence used in the packaging of food products.

The plastic industry is divided into four categories – bags, household and kitchen wares, industrial plastic supplies and miscellaneous items. The industrial plastic supplies cover items such as casings for radios, cassette recorders and TV sets; and PVC pipes and fittings for the building industry.

The miscellaneous items include precision-molding of plastics to produce gears and bobbins for digital clocks; and rollers, hubs and clamps for audio and video cassettes.

The plastics industry has achieved remarkable success by widening the use of its products as well as by diversifying its applications.

Metallurgy

Metallurgy is the science of extracting metals from their ores and preparing them for practical and commercial use. Metallurgical processes include the refining of metals and the manufacture of alloys for special purposes. The manufacture of steel is especially important because of its wide application.

Pharmaceuticals

Since ancient times, we knew that some substances had the power to heal certain diseases. Many plant extracts had such medicinal properties. In the beginning, chemists concentrated their attention on isolating the active components in such extracts and using them. Nowadays, chemists are involved in the synthesis of such compounds. The discoveries of sulphonamides and antibiotics have led to a great improvement in the quest for a better life. Drugs of today have conquered many diseases considered incurable in the past. They include:
- quinine for the treatment of malaria, and
- insulin for the treatment of diabetes.

The majority of the drugs manufactured nowadays are synthetic. Some of them are identical to natural substances found in plants and animals, while others are entirely new. Especially useful are the sulpha drugs which are used to treat pneumonia and other serious illnesses. The most widely used drug is aspirin. It was discovered in coal tar, but nowadays it is manufactured synthetically.

Glass

Sand appropriately blended with various other substances has given us glass. Window panes, mirrors, bottles, fruit jars, drinking glasses and dishes are made of glass. Lenses for still and television cameras are made of glass. Chemists use a large number of test-tubes, beakers and other apparatus made of glass for their experiments. There are many kinds of glass. They may be transparent, translucent or colored, depending on the purpose for which they are made. Glass used in ovenware has to be heat resistant. The glass used in test-tubes and beakers must be resistant to chemicals. To produce these different properties, the composition of glass and the heat treatment have to be varied. In nature, glass is formed during volcanic eruptions. The Egyptians were the pioneers in the art of making glass, but the foremost glassmakers of the olden days were the Romans. Today, the industry is almost completely mechanized, though hand operations are still used in some small-volume specialized fields. The chief processes used for making glass are much the same as they have been in the past. Sand, soda, lime and other materials are mixed together and melted in a furnace at a very high temperature to produce molten glass. While the glass is molten, it is shaped and made to cool so as to form a rigid piece. The glass piece is annealed by reheating it and then gradually cooling it.

In modern glass manufacturing factories, machines do a lot of work that were done by hand previously. Ovenware, heavy tumblers and many other kinds of glassware are shaped entirely by pressing. A quantity of hot glass of the right size and shape is delivered to a metal mould which is set in position under a plunger. The plunger forces the hot glass into the space between the mould and the plunger and holds it there until it has solidified.

Window glass is made by drawing a continuous vertical sheet from a pool of molten glass. Plate glass is made by rolling.

Massive shapes such as architectural blocks and mirror banks used for big telescopes are made by casting. For this purpose, heat-resistant moulds are used and molten glass is poured into them.

Like most materials, glass contracts on cooling. Furthermore, glass is a poor conductor of heat. Thus, it is quite possible that during the forming process, various parts of the glass may be at different temperatures and will solidify at different times. Each part will tend to contract by different amounts as it cools to room temperature. Since cold glass is rigid, the difference in contraction will cause internal stress to develop. To prevent this, the glass must be annealed. Most glass today is annealed in continuous furnaces called *lehrs*. The glassware is placed on a metal belt that conveys it through a tunnel. The temperature along the tunnel is graduated so that after passing through a hot zone, the glassware is slowly cooled to a safe handling temperature at the end of the belt.

Ceramics

The art of processing earthy materials into useful or ornamental objects through the application of high heat is known as ceramics. The art of making ceramic products goes back beyond the era of recorded history. Up to the early years of the twentieth century, ceramics were produced by trial and error. In recent years, chemistry and physics have joined hands to unravel the secrets of ceramic materials. This has resulted in better products and more varied uses. Scientists can now produce ceramics that can withstand the stress and heat imposed by the supersonic speeds of aircraft and by nuclear reactors. They are also used in critical parts of rockets and spacecraft, such as the nose cones, rocket exhaust nozzles and heat-resistant windows.

The raw materials used are silica and oxides of magnesium, aluminum, zirconium, thorium, titanium and boron. Some artificial carbides and nitrides are also used nowadays. The raw materials are ground to a uniform consistency and mineral impurities are removed. The cleaned materials are then heated to temperatures between $500°C$ and $3000°C$. During these processes, certain physical and chemical changes are brought about. These changes make the ceramics strong and durable.

Paints

Paint is a fluid mixture which contains suspended coloring material. When it is applied as a thin layer on a surface, it produces an opaque solid film. Although its early use was chiefly decorative, protection against weathering and corrosion is an equally important use in the modern times.

Most paints are applied by brushing, but other methods such as spraying, dipping and roller coating are also used.

Cement

Cement is made by heating a mixture of powdered lime (calcium oxide) and clay. When mixed with water, it can be used to fasten stones and bricks together. The mixture (called mortar) hardens like stone when it dries. The most popular cement is known as Portland cement. This type of cement is made by heating a mixture of limestone and clay, and grinding the products. It consists of calcium silicates (as a result of the reaction between silica and calcium oxide) and calcium aluminates (formed by the reaction between calcium oxide and aluminum oxide present in clay). Complex chemical changes occur during setting, resulting in a hard mass. Portland cement is mainly used as a component of concrete. Concrete is made by mixing cement, sand and broken stones or gravel. The sand and gravel must be clean and free from soft particles and vegetable matter. Water is then added to the mixture which forms a rock-like mass when it hardens. In the late nineteenth century, it was found that concrete can be reinforced with steel rods. Reinforced concrete revolutionized the building industry. It is used in highways bridges and dams. It is also used in virtually all large buildings. One of the advantages of concrete is that it can be poured into forms, and it hardens in place so that the whole structure is like one huge stone. There are no joints to worry about.

Soaps and Detergents

Soap is defined as a chemical compound or a mixture of chemical compounds resulting from the reaction between fatty oils or fats, and alkalis. It is manufactured by introducing melted fats into an excess of sodium hydroxide solution and boiling the mixture. The pasty boiling mixture is then treated with brine. This causes the contents to separate into two layers. The upper layer contains impure soap. This is then washed to remove the excess alkali and salt, and treated in various ways to give the desired product. Detergents are synthetic cleansing products. They are often made from petrochemicals. Detergents are used to wash dishes, laundry, walls and floors. Soaps and detergents have the property of reducing the surface tension of water. They are able to lift dirt from greasy surfaces thereby reducing the spread of germs.

Food and Drinks

Chemical and biochemical processes are used in preserving, flavoring and enriching the nutrient content of food and drinks. The production of alcoholic drinks such as beer involves fermentation processes, which have to be carefully controlled. In the preparation of food and drinks, certain specified standards have to be met. Chemists analyze samples of food and drinks routinely to ensure that these standards are met.

Economics of Industrial Processes

The economics of an industrial process is mainly linked to:

- raw materials,
- energy requirements
- transport
- use of by-products and recycling, and
- supply and demand.

Raw Materials, Energy Requirements and Transport

Raw materials should be easily and cheaply obtainable. Fuel supplies are very expensive. So the energy requirement of an industrial process has to be taken into consideration before starting a particular chemical industry. Road, rail and sea links should be present to transport both raw materials and finished products. These three factors should be taken into account in the citing of a chemical factory. Only then can the industrial process be economically feasible.

Use of by-products and recycling

Where possible any chemical produced as a by-product should be used so that it does not form waste. An industrial process should also ensure used products and waste materials can be returned to the processing stage, i.e. recycled. Recycling conserves the world's supply of non-renewable raw materials such as metals. It is also economical as the recycled material does not pass through the energy demanding extraction stage.

Supply and demand

The demand for a certain substance may change suddenly so that the basic process of a chemical industry has to be altered for it to be economical. For example, in the past, the heavy fractions of crude oil were in great demand. Now, the lighter petrol and natural gas fractions are in great demand. Crude oil industry is meeting this demand by cracking chemicals in the heavier fractions to produce the smaller molecules found in the lighter fractions.

Importance of Industrial Chemistry

Chemical industries require:

- raw materials,
- energy,
- water,
- machinery,
- labour, both skilled and unskilled,
- transportation for the raw materials and products, and
- markets for the products.

My countries have a rich supply of petroleum, natural gas and coal which are important raw materials needed in chemical industries. Some other important inorganic raw materials include limestone, tin, columbite, clay and iron ore. Common raw materials of plant and animal origins include rubber, palm oil, oils from groundnut and cotton seed, and cellulose. Many countries use these raw materials and many more for their chemical industries. They also exports raw materials to countries which need them for their own industries. Chemical industries require energy. The amount and type of energy needed will depend on the chemical process. Some countries have their own supply of energy from fossil fuels like petroleum, natural gas and coal, and also from moving water (hydroelectric power). The power plants which supply electrical energy to the chemical industries use mainly petroleum and natural gas. There are factory which use coal as major source of energy. Most chemical industries need plenty of water, especially for cooling. Very pure water is needed for certain chemical processes like dyeing. Some developing nations have to spend money to import machineries that are needed in chemical industries.

Chemical industries need highly skilled chemists, chemical engineers and technicians. Our schools, colleges and universities must have the facilities to train such people. The requirement for unskilled labour is not high in chemical industries as in other industries like building and food processing industries. Adequate road and railway transportation are necessary. Industries which use heavy raw materials like cement and iron ore are usually located in the areas where the raw materials are produced. Chemical industries that use imported raw materials or which export a greater part of their products are built near coastal towns and ports. Countries export both raw materials for chemical industries and products from their own chemical industries to other countries. The chemical industries also supply many products for domestic use. Thus, we see that chemical industries:

- provide earnings from foreign trade;
- improve the standard of living by providing many materials for domestic use (which would be too expensive if they were imported); and

- provide employment.

Many chemical industries are carried out on a large scale. There are also small scale chemical industries such as pottery, dyeing and tanning of leather.

Cleaning up Wastes

Most chemical wastes produced by chemical industries are harmful to man and they pollute our environment. Such wastes include:

- radioactivity wastes from nuclear reactors,
- sewage from brewery or soap and detergent industries,
- slag from iron and steel industries,
- paints,
- solvents,
- cyanide,
- polychlorinated biphenyls,
- lead,
- mercury and
- banned pesticides.

These wastes are usually dumped in rivers, sea, or land and they constitute health hazards. Cleaning up a hazardous waste site is a complete project. One project may bring together experts from the field of:

- chemistry,
- biology,
- hydrogeology,
- law and
- politics.

The experts will study the site and decide upon the best cleanup approach to take. Traditional cleanup methods include:

- incineration,
- physical containment and
- land burial.

Bioremediation is a new method currently being developed to destroy toxic wastes. Bioremediation uses bacteria that naturally degrade hazardous materials. The cleanup site is sprayed with special fertilizers that encourage bacterial growth. The degradation process can be accelerated up to one million times the rate of normal degradation. Bioremediation has been shown to be effective in treating city and company sewage.

In order to see other mathematics, physics and chemistry books written by the author, visit: amazon.com/author/kingzohb2. Also, you can simply go to amazon.com and search for the author's name, Kingsley Augustine, the books written by the author will show up.

If you have any enquiries, suggestions or information concerning this book, please contact the author through the email below.

Kingsley Augustine

kingzohb2@yahoo.com

www.ingramcontent.com/pod-product-compliance
Lightning Source LLC
Chambersburg PA
CBHW071019290526
45795CB00005B/1869